I0844829

# The Climate Catastrophe:

# Earth's Battle for Survival

## By

## Steven Hannan

# TABLE OF CONTENTS

## Introduction

The globe is currently engaged in a terrifying struggle for survival in a time marked by unheard-of environmental problems.

The global disaster known as the "Climate Catastrophe," which endangers the structure of our planet itself, looms menacingly on the horizon.

As the Earth's ecosystems, cultures, and species face an existential peril unlike any other, this apocalyptic conflict cuts over national boundaries, political views, and generational divides.

In this investigation, we set out on a quest to comprehend the nuances of this problem, its origins, and the urgent actions needed to lessen its disastrous repercussions. Welcome to the front lines of the fight for Earth's survival, where the future of our planet is at stake.

**Part I:**

**Recognizing the Crisis**

***The Gathering Storm: Worries About Climate Change Are Growing**

Climate change puts our planet in ever-greater peril as the world deals with a growing number of environmental issues.

The primary causes of this problem, which include burning fossil fuels and deforestation, are human actions.

As a result, there have been several unintended effects that are presently escalating alarmingly. We will address how climate change has changed and intensified as well as its wide-ranging effects on our environment, society, and the future.

Climate patterns have changed in recent years, which is one of the most striking changes.

Unprecedented increases in global temperature are causing more frequent and severe heat waves. Ecosystems are disrupted, water resources are strained, and agriculture is threatened as a result of these changes, which eventually undermine global food security.

A rise in extreme weather occurrences, from protracted droughts to strong rainfall and flooding, has also been linked to changing precipitation patterns.

An increase in sea levels Rising sea levels are a result of glaciers and polar ice caps melting more quickly due to warming temperatures. Millions of people could be displaced as a result of the inundation of their houses and essential infrastructure in coastal regions, which are especially susceptible to this phenomenon.

The economic effects of this altered climate hazard are severe, and coastal communities will incur high adaptation costs.

An increase in natural disasters is caused by climate change. Extreme occurrences including hurricanes, typhoons, wildfires, and other catastrophes have increased in frequency and severity. The threat to life and property is grave, and this escalation puts a burden on emergency response services.

Communities all across the world need to construct resilient infrastructure to prepare for these heightened dangers.

Many ecosystems are on the verge of extinction due to altered climate conditions. Forests are becoming more vulnerable to pests and wildfires, while coral reefs are experiencing episodes of bleaching. Accelerating biodiversity loss disturbs sensitive biological balances and lowers the planet's overall resistance to environmental shocks.

Accelerating climate change is creating a storm that affects not only the natural world but also society and the economy. The effects of climate-related food shortages and displacement are mostly felt by vulnerable groups, who frequently contribute the least to greenhouse gas emissions.

Economic costs from climate-related calamities are increasing, endangering lives and putting the world's stability in jeopardy.

Two strategies are needed to address these modified climate challenges. Transitioning to renewable energy sources, improving energy efficiency, and replanting all play crucial roles in mitigation efforts, which are necessary to minimize greenhouse gas emissions.

In addition, it is essential to implement adaptation techniques to help communities get ready for the unavoidable impacts of climate change, such as building resilient infrastructure, being prepared for emergencies, and implementing fair regulations. The storm that is speeding up climate change is a serious worldwide issue that has to be addressed right away. Wide-ranging societal and economic effects, altered climatic patterns, increasing sea levels, intensifying natural disasters, ecosystem collapse, and other factors all call for comprehensive solutions. The only way we have any chance of preventing the most dire effects of this rising threat from climate change and guaranteeing a sustainable future for future generations is via concerted efforts to reduce emissions and adapt to the new reality.

## *Identifying the Key Drivers of Climate Change"

The serious global challenge of climate change calls for in-depth comprehension and immediate action. *The ground-breaking investigation "Gathering Storm Unmasking the Culprits"* takes a close look at the primary causes of this environmental calamity.

The documentary series Gathering Storm tries to expose the covert causes of climate change. We must determine the main causes of the warming of the globe to completely understand this complex topic.

The greenhouse effect is one of the main causes of climate change. *Carbon dioxide (CO2), methane (CH4), and nitrous oxide (N2O)* are examples of greenhouse gasses that trap heat in the Earth's atmosphere and cause an increase in global temperatures. The origins, effects, and necessity of emissions reduction will all be included in this modified content.

Deforestation and irresponsible land use are two more major factors. In addition to reducing carbon sinks, the destruction of forests for agriculture and urbanization also releases stored carbon into the sky. We'll talk about how sustainable land management and replanting might lessen this problem.

Although the Industrial Revolution resulted in significant breakthroughs, our environment paid a price. *CO2* and other pollutants are released when fossil fuels like coal, oil, and natural gas are burned. The significance of switching to clean and renewable energy sources will be emphasized in this section. How we produce and consume our food also has a big impact on climate change. This section will examine how agricultural methods, such as monoculture and animal farming, affect greenhouse gas emissions and deforestation.

Dietary options and sustainable agriculture will be covered as potential remedies. To effectively prevent climate change, we must address both the causes and the effects. Meaningful change can be sparked by ***governmental initiatives, international collaboration, and technical advancements***. We'll look at how these elements contribute to climate change mitigation in this section. The Gathering Storm initiative demonstrates how urgent climate change action is. Inaction has disastrous effects that range from increased frequency and severity of natural disasters to the uprooting of entire communities.

We will stress the necessity of swift, international cooperation in resolving this problem. Unmasking the Culprits" illuminates the primary causes of climate change and calls on society to shoulder its share of the blame. We can increase awareness of these offenders and encourage action to protect our world for future generations by altering the way we produce information.

## *Earth's Vital Signs: Tracking the Health of Our Planet.

an essential and timely issue that illustrates our growing environmental concern. A comprehensive monitoring system is necessary to comprehend the many factors that affect the health of our planet and how they are changing over time.

One of the most crucial parts of the system is the gathering of data regarding climate change. Earth's temperature is rising as a result of greenhouse gas emissions, leading to more frequent and severe weather events, rising sea levels, and changes in ecosystems.

By keeping an eye on these temperature swings in conjunction with data on carbon dioxide levels and other greenhouse gas concentrations, we may better comprehend the impact of human activity on our climate.

On Earth, the state of the oceans must also be continuously monitored. Ocean acidification, rising ocean temperatures, and plastic pollution are all major threats to marine life.

Understanding these changes can help conservation efforts and legislation be directed toward maintaining these crucial ecosystems. The system also keeps track of biodiversity.

As species experience issues due to climatic change and habitat loss, monitoring ecosystem changes might show early warning signs of looming catastrophes.

This knowledge is crucial for conservationists working to preserve biodiversity and save endangered species. Furthermore, one of the most important indications for the planet is the state of our forests. Deforestation and forest degradation have an influence on carbon emissions as well as the loss of habitat for some species. By monitoring these developments, we might design plans for sustainable forestry and reforestation initiatives. Water resources are still another crucial element. ***Droughts, pollution, and water shortages all harm ecosystems and societies***.

Monitoring these variables aids in developing emergency plans and water management strategies. *Fundamentally, monitoring and assessing the planet's major indicators serves as a barometer for its health.* With the help of this knowledge, scientists, policymakers, and regular people can make informed decisions that will mitigate the effects of climate change, protect biodiversity, and ensure a sustainable future for future generations. It stresses how urgent it is for nations to work together to address the issues that are endangering the planet's health.

## Part II:

## The Frontlines

## *Melting Ice Caps & Sea Levels

The subject of climate change must include the melting of the polar ice caps and rising sea levels. Understanding how these regions interact is essential because of how profoundly the effects of climate change have affected them. Because of the increase in global temperatures, the Arctic and Antarctic polar ice caps are melting at an alarming rate. The melting of this ice, which holds a sizable percentage of the freshwater on Earth, causes the sea level to increase. There are numerous effects of this phenomenon.

First of all, coastal communities all around the world are in danger due to rising sea levels. Coastal areas are more susceptible to floods and erosion as ice melts and seawater expands as a result of rising temperatures.

This affects ecosystems, infrastructure, and residences, forcing millions of people to leave their homes and resulting in financial losses.

Additionally, the melting of ice caps has an impact on weather patterns worldwide. The equilibrium of heat in the atmosphere and oceans is upset by melting ice, which causes unpredictable weather conditions such as stronger storms, droughts, and heat waves.

The effects of these shifts on agriculture, water availability, and overall climatic stability are grave.

The effects of ice cap melting and rising sea levels are not just felt in the nearby coastal areas. Low-lying island nations confront the existential threat of becoming uninhabitable as sea levels continue to rise. Global difficulties can also be made worse by the displaced populations caused by flooding and shifting weather patterns, which can lead to migration and conflict. Reducing greenhouse gas emissions, switching to renewable energy sources, and putting adaptation measures in place in susceptible places are all necessary for mitigating these effects.

It is impossible to exaggerate the importance of addressing climate change, as the health of the Polar Ice Caps and the effect they have on sea level rise are emblematic of the larger global problem that impacts us all.

***Conflicts over global water resources, scarcity, pollution, and water wars are all related to climate change.**

The foreboding combination of water scarcity, pollution, and the possibility of conflicts over this priceless resource, all of which are compounded by the threat of climate change, has become an increasingly worldwide issue in the twenty-first century. It becomes clear as we go more into this intricate web of problems that they are intertwined in a way that necessitates quick attention and creative solutions.

The depletion of freshwater supplies is one of today's most urgent issues. By changing precipitation patterns, climate change has made this issue worse by causing disastrous floods in some areas and lengthy droughts in others. The need for water, mostly for domestic, industrial, and agricultural usage, is only increasing as the world's population grows. Water scarcity has already led to tensions and strained international relations in places like the Middle East.

A comprehensive strategy that incorporates equitable distribution, conservation, and sustainable water management is required to address this issue.

Chemical and biological pollution seriously endangers the availability of freshwater worldwide.

Freshwater sources are contaminated, making them inappropriate for human consumption and detrimental to the health of the ecosystem.

These contaminants come from industrial effluents, agricultural runoff, and untreated sewage. By raising water temperatures, which can encourage the growth of hazardous microbes and produce dead zones in bodies of water, climate change can exacerbate this problem.

To alleviate this situation, effective pollution control strategies and the promotion of environmentally friendly behaviors are needed.

Conflicts over water rights and access are more likely as freshwater supplies grow more scarce and contaminated. Water issues have historically caused friction and sometimes war between nations. The topography of water availability has changed as a result of climate change, further complicating this terrain.

Long dry spells could affect areas that formerly depended on regular rainfall, increasing competition for few resources. Treaties, international collaboration, and processes for resolving disputes are essential for preventing water wars and advancing peaceful solutions.

***The impact of global warming:*** Water scarcity and pollution are exacerbated by climate change, which works as a multiplier.

Water supplies become increasingly unpredictable as a result of the disruption of regular weather patterns. Evaporation rates may increase with a temperature rise, severely depleting freshwater supplies.

In addition, severe weather conditions like hurricanes and cyclones can contaminate water supplies.

Climate change mitigation and adaptation measures must be combined with water resource management initiatives to effectively address these issues.

Water scarcity, pollution, and the possibility of international water disputes are three severe realities that require immediate response. The issues are made more difficult by the effects of climate change.

International collaboration, sustainable water management techniques, pollution control, and climate change mitigation measures must work together to traverse these choppy waters. Our ability to conserve and maintain this priceless resource while promoting peace among nations dealing with this common challenge is crucial to the future of our planet.

**Part III:**

**The Human Component**

***Mass migration in a world in transition: Climate Refugees**

Climate refugees, sometimes referred to as environmental migrants or climate-displaced people, are becoming an increasingly serious global issue as the effects of climate change worsen.

This phenomenon is a modified result of the society we live in today, as people are being forced to leave their homes due to climate change-related factors like sea level rise, harsh weather, and rising temperatures.

The importance of this issue's scope should be one of its main topics. By the middle of the century, hundreds of millions of people may have been uprooted due to climate change, according to the Intergovernmental Panel on Climate Change (IPCC).

Both those affected by and the receiving communities are faced with formidable obstacles as a result of this huge migration.

The reasons for climate-induced relocation are a crucial issue that must be addressed.

For instance, long-lasting droughts have the potential to progressively render an area uninhabitable, whereas abrupt disasters like hurricanes or floods might compel quick evacuations. Knowing how these occurrences affect human mobility is essential because of how closely these phenomena are related to climate change. It is crucial to talk about how migration is being caused by the climate. Conflicts over scarce resources, congested cities, and employment opportunities might result from it. Furthermore, it calls into question the moral obligations of developed nations towards climate refugees and the necessity of international cooperation to address this problem.

As for solutions, accommodating climate refugees will require a revised approach to immigration laws and humanitarian aid.

The creation of a legal framework that recognizes climate displacement as a legitimate justification for requesting asylum or relocation requires cooperation between governments and international organizations.

Additionally, addressing the underlying causes of climate migration requires both mitigation of the effects of climate change and adaptation to those effects. We can limit the rate of environmental deterioration and the number of people compelled to leave their homes by reducing greenhouse gas emissions and putting sustainable practices into practice. Our planet is changing, and there are more and more climate refugees. Understanding the causes, consequences, and potential solutions to this issue is crucial for a sustainable and compassionate response to the plight of those affected by climate-induced displacement.

## *The Function of Big Business in Combating Corporate Climate Change Influence

The grave global problem of climate change calls for coordinated action. Although individuals, groups, and governments all have a function to play, it is important to not undervalue the influence of large industries.

Because of the size of their operations and resource requirements, large enterprises have a major negative influence on the environment. This impact has the potential to either exacerbate the climate problem or be used to spur progress.

Many large industries have historically come under fire for putting profit ahead of environmental concerns. For instance, companies that use fossil fuels have significantly increased greenhouse gas emissions. Beyond their direct emissions, they also frequently lobby against climate laws and support efforts that deny the existence of global warming.

The world's efforts to address climate change have been hampered by corporate influence. However, there has been a change in how corporate responsibility is perceived in recent years.

The need to address climate change has been acknowledged by many significant businesses, not only for moral reasons but also to protect their long-term sustainability.

Some people have committed significantly to reducing their carbon footprint and switching to renewable energy. Growing customer demand for environmentally friendly goods and services is one of the factors driving this transition.

Transparency is a crucial component of addressing corporate influence on climate change. Greater responsibility from large industries is being demanded by the general public and stakeholders. For many businesses, reporting on emissions, sustainability objectives, and progress toward these objectives has become normal procedure. Through increased transparency, customers can make better decisions, and businesses are held more responsible.

Governments are also essential in regulating big business to conform to climate goals. Setting and implementing carbon targets, encouraging green investments, and enforcing fines for non-compliance are all necessary to achieve this.

By establishing challenging goals for lowering emissions, *international agreements like the Paris Agreement have strengthened the global movement for corporate accountability.* Another viable path is cooperation between large businesses and environmental NGOs. In addition to influencing public opinion and governmental regulations, partnerships can foster innovation in sustainable behaviors and technologies. Even some businesses have embraced the ideals of the circular economy, which aims to reduce waste and increase resource efficiency.

In the fight against climate change, big business is crucial. Although corporate influence has traditionally been a hindrance, the evolving environment implies that it can also be a driver for good. Utilizing the capacity of large businesses to fight climate change requires making the shift to sustainable practices, enhancing transparency, and working with other stakeholders. Making sure business interests align with the global duty to mitigate climate change for the benefit of the earth and future generations is the task that lies ahead.

# Part IV:

# Hope and Solutions

# *Green Technology Innovations: Breakthrough Sustainable Solutions

Green technology advancements have been crucial in tackling environmental issues and promoting sustainability in a variety of industries.

These ground-breaking approaches incorporate a variety of innovations meant to cut carbon emissions, save resources, and encourage a future that is cleaner and more sustainable. Renewable energy is one notable area of innovation. It is now simpler for people and companies to harvest clean energy from renewable sources because of improvements in solar panels and wind turbine efficiency and cost.

The dependability of renewable energy systems has also increased thanks to energy storage technologies like cutting-edge batteries, which guarantee a steady supply of electricity even when the sun isn't shining or the wind isn't blowing. The field of transportation is home to yet another potential breakthrough.

Due to their smaller carbon footprint and cheaper operating expenses when compared to conventional internal combustion engine vehicles, electric vehicles (EVs) have become increasingly popular.

The range of EVs is increasing and their charging times are getting shorter because of ongoing development in battery technology, which also makes them more affordable for customers. Innovations in green technologies also affect the built environment. *Buildings that are smart and save energy today come with sensors and automation systems that optimize energy use and raise sustainability standards.* Infrastructure projects now have less of an environmental impact thanks to advancements in materials science that have produced eco-friendly building materials and methods. Precision farming techniques in agriculture are assisting farmers in maximizing crop yields while reducing resource inputs like water and fertilizer. Traditional agricultural techniques are being reimagined by vertical farming and aquaponics, enabling year-round, sustainable food production in urban environments. In addition, the idea of the circular economy, which emphasizes the value of minimizing waste and optimizing resource reuse and recycling, is gaining popularity. More materials are being diverted from landfills because of advances in waste management and recycling technologies, which also help save resources and lessen pollution. These are only a few instances of how advancements in green technology are laying the foundation for a more sustainable future.

We may anticipate even more ground-breaking innovations that will further alleviate environmental effects and advance a greener, more sustainable future as technology develops.

## *Grassroots Initiatives and Climate Activism: A Global Movement

A notable increase in grassroots initiatives and global climate advocacy has been seen in recent years. This influential movement has evolved in reaction to the urgent environmental issues that endanger the future of our world.

We will analyze the importance of these initiatives, how they have changed over time, and how they have contributed to long-lasting change in this debate. In the context of climate action, grassroots initiatives are community-led projects that seek to address environmental challenges locally. They frequently start with concerned individuals or small activist organizations who are aware of the urgency of climate change and its severe repercussions.

These programs come in a variety of shapes, from neighborhood-based renewable energy projects to cleanups of plastic waste and tree-planting campaigns.

One famous instance is **_Greta Thunberg's_** **_"Fridays for Future" movement_**, which inspired global climate strikes and inspired young activists all over the world. The flexibility and possibility for adjustment of grassroots efforts is what makes them so beautiful.

They can change and adapt to new situations and obstacles as they arise.

These programs have integrated creative solutions and modified their strategy as our awareness of climate change grows and new technology is developed.

For instance, community-owned solar and wind energy projects have emerged as a result of the spread of renewable energy technology, enabling nearby communities to cut their carbon footprint and achieve energy independence.

In addition to winning over people's hearts and minds, these grassroots projects have also attracted a lot of interest from governments, corporations, and international organizations. As a result, they have had a significant impact on corporate sustainability initiatives and climate legislation. These movements have put pressure on governments to establish more aggressive climate goals and businesses to adopt greener practices.

Furthermore, grassroots initiatives have shown the ability to have an impact over the long term. They bring about long-lasting change that goes above and beyond the short-term objectives of particular projects by encouraging a sense of community ownership and involvement. A cultural shift towards environmental stewardship is facilitated by this grassroots-driven strategy, which promotes people to make sustainable decisions in their daily lives and increases awareness of the significance of taking action on climate change.

A ray of hope in the fight against climate change is the global movement of grassroots projects and climate activism. These projects are important participants in the group effort to address the climate problem because of their adaptability, capacity to involve various people, and ability to influence policies.

They have the power to bring about the transformational change required to ensure a sustainable and resilient future for our planet as they develop and inspire others.

## *Approaches to a Sustainable Future of Climate Change

A multifaceted strategy is needed to build a sustainable future in the face of climate change.

*Key tactics are as follows:*

*Reduce greenhouse gas emissions by switching to renewable energy sources like solar, wind, and hydropower instead of fossil fuels.

***Energy Efficiency**: Through improved technology and practices, increase the energy efficiency of buildings, transportation, and industry.*

***Carbon Pricing:** Use carbon pricing tools like carbon taxes or cap-and-trade programs to encourage the reduction of emissions.*

*to absorb carbon dioxide from the atmosphere, reforestation, and afforestation include expanding and restoring forests.

*Promote environmentally friendly farming methods, maintain healthy soil, and save biodiversity.

*Accelerate the adoption of electric automobiles and spend money on public transit.

***Circular Economy:** Move in the direction of a circular economy, which minimizes waste and encourages recycling and reusability.*

*The people should be informed about climate change and sustainability to make wise decisions.

*Building infrastructure and implementing policies to adapt to a changing environment *and reduce hazards is known as resilience planning.*

***International cooperation:*** Work together to establish bold climate goals and agreements.

***Corporate responsibility:*** Promote the use of sustainable business practices and the lowering of carbon emissions.

*Encourage people to take individual actions to lessen their carbon footprint by using less energy, generating less waste, and shopping sustainably.*

***Technological Innovation***: *Fund the study and creation of eco-friendly technology.*

***Policy and Regulation:** Implement and uphold laws and rules that encourage sustainability and punish non-sustainable behavior.

*Ecosystems that trap and store carbon, such as wetlands and mangroves, should be preserved and restored.

***Climate adaptation:** Create plans for communities to deal with the unavoidable consequences of climate change.

*Address the disproportionate effects of climate change on disadvantaged groups as it relates to environmental justice.

***Long-Term Planning:** Implement long-term plans to make sure sustainability transcends passing political trends.

*__Consumer Awareness__: Encourage responsible consumption and support eco-friendly products and services.

*__Financial Incentives:__ Offer financial rewards for spending money on sustainable activities and clean energy.

Collectively, these tactics can slow down climate change and contribute to a more sustainable future.

## In Conclusion

Mobilizing for a Climate-Resilient World" highlights how quickly climate change needs to be addressed. This debate will look at how we might alter our behavior and come up with long-lasting fixes to create a world that is climate resilient.

Our planet is in peril from climate change, which affects ecosystems, economy, and society. We need to change our course if we're going to address this catastrophe. This begins with lowering greenhouse gas emissions through the use of clean energy sources, sustainable farming methods, and energy-saving devices.

Long-term solutions must be created using a comprehensive strategy. To create policies and innovations that support climate resilience, governments, corporations, and individuals must work together.

This includes making investments in infrastructure for renewable energy sources, safeguarding natural habitats, and improving preparedness for disasters.

In addition, spreading knowledge and awareness is essential for creating long-lasting change.

Future generations must be taught the value of environmental preservation and sustainable life.

We can cultivate a culture of sustainability that will last for years to come by changing our actions and supporting green activities.

To battle climate change, "Mobilizing for a Climate-Resilient World" urges urgent action, changes to our habits, and the creation of all-encompassing, long-term solutions.

We can only guarantee a sustainable and resilient future for our world by working together.

***The Climate Catastrophe***" is intended to inform readers, spur action, and serve as a crucial tool in the effort to protect the future of our world. Together, we can change the course of the struggle for Earth's existence.

www.ingramcontent.com/pod-product-compliance
Lightning Source LLC
Chambersburg PA
CBHW071100260726
48661CB00006B/2360